PROSPECTUS

FOR THE FORMATION OF THE

Lake Superior Silver-Lead Co.

Sections 6 and 36, T. 49 and 50, R. 28 and 29,

MARQUETTE COUNTY, LAKE SUPERIOR, MICHIGAN.

NEW YORK:
BAKER & GODWIN, PRINTERS,
PRINTING-HOUSE SQUARE, OPPOSITE CITY HALL.
1863.

PROSPECTUS

FOR THE FORMATION OF THE

Lake Superior Silver-Lead Co.

Sections 6 and 36, T. 49 and 50, R. 28 and 29,

MARQUETTE COUNTY, LAKE SUPERIOR, MICHIGAN.

NEW YORK:
BAKER & GODWIN, PRINTERS,
PRINTING-HOUSE SQUARE, OPPOSITE CITY HALL.
1863.

PROSPECTUS

FOR THE FORMATION OF THE

Lake Superior Silver-Lead Company,

SECTIONS 6 & 36, T. 49 & 50, R. 28 & 29.

Marquette County, Lake Superior, Michigan.

THE proprietors of the above-named lands propose to organize a Company with the title of the LAKE SUPERIOR SILVER-LEAD MINING COMPANY, to work the valuable veins of argentiferous-lead ore known to occur on them.

The property consists of ten hundred and thirty acres of mineral land, in the County of Marquette, State of Michigan. On these, several lodes of silver-lead ore have been recently discovered. These, it is well established, are of unprecedented size and richness in silver, and one of them has been traced for miles across the country in a northwest and southeast direction, being a well-marked fissure vein, and, from its great size and extent, cannot but be regarded as a champion or master lode. It is contained in rocks of the most favorable geological age to be productive in silver; and assays show that its vein-matter is no exception in this respect to the general rule, unless it be in its extraordinary richness. It is eight feet in width, and runs through the property for a length of fully three miles, giving abundance of ground for extensive operations. The galena is contained in a gangue of quartz, in which it is plainly developed and thickly disseminated.

In addition to this amount of land, there are one hundred and thirty-five acres on Big Bay, twelve miles north from the main body of lands on Lake Superior, and seventy-eight acres on Huron Bay. These last two mentioned tracts have been secured for the purpose of having an outlet or "embarcadero" for the shipment of the products of the mine, and a depot for the reception of supplies, etc.

Since the establishment of the existence of rich silver-bearing lead ores, the district has been visited and explored by many of the best known geologists and mining men on the Lake, who have expressed themselves in the highest terms with regard to its value, having satisfied themselves of the existence of heavy bodies of the ores, and of the prospects of opening on these exceedingly profitable mines by the investment of but a comparatively small amount of capital.

The extreme richness of the ores from this new mining region of Lake Superior can be better illustrated by the comparison of their yields with those from some of the best known and most profitable mines of silver ores proper and of argentiferous galena in other countries.

In 1852, the following were the yields of silver from the silver-lead mines of Great Britain.*

* Robert Hunt in Ib., p. 367.

Oz. of Silver per ton of Lead.		*Oz. of Silver per ton of Lead.*	
Cornwall,	35	Cardigan, Caernarvon, and Caermarthenshire,	15
Devon,	40	Montgomery and Merionethshire,	6
Durham, Northumberland and Westmoreland,	12	Ireland,	10
Cumberland,	9	Scotland,	8
Flintshire and Derbyshire,	7	Isle of Man	20

The lead-bearing veins of the Harz carry silver in amount ranging between 13 and 123 ounces per ton of dressed ore. Many of these mines have been worked for centuries, and the produce of some of them has been remarkably steady for many years. The age of the geological formation in which the deposits occur is the same as that of the Lake Superior region. In Nassau, at Holzappel, silver-lead has been mined since 1158. The yield is said to be about 75 ounces to the ton of dressed ore.

Among the mines of Saxony, that of the Himmelfahrt ranks first in production, furnishing, from 1830 to 1850, 146,824 lbs. troy of silver, and paying a profit in ratio to that quantity of metal of 1 to 22.35. In the ten years ending 1861, it paid, in the shape of dividends, $1,250 per share; and in the year 1862 it paid $450 per share. The assets of the Company and the cash balances have constantly increased, and at the close of 1861, the assets amounted to upward of $535,545, the reserved fund being equal to $1,915 per share.

As an instance of what can be done with only a small tenure of silver in the vein-stuff, the Mansfield mines of Germany may be cited, though not properly belonging to the subject of silver-lead. The deposit is only from four to six inches in thickness, and the amount of silver contained in the undressed ore is one-seventieth of one per cent. The present weekly yields exceed $40,000 worth of silver, and $80,000 worth of copper. The dividends paid for the five years ending 1859 amounted to $1,446,425. The ore occurs in a remarkable bituminous schist, in the Permian group of the English geologists.

The mine at Konigsberg is opened on veins contained in slates of the azoic age. It was commenced as early as 1624, and has been wrought, with some few short interruptions, up to the present time. From that time to the close of 1838, its production was 1,733,-879 lbs. troy of silver. From 1849 to 1853, its average annual produce was 16,971 lbs. at a yearly profit of $160,000 (average.) At the present time, the average annual value of the silver is $275,550; its average annual expenses $54,550; and its yearly profit is $221,000.*

The mine of Bronfloyd in Wales, paid in October last its usual dividend of twenty-two per cent. per annum. The amount of silver is not greater than fourteen ounces per ton of dressed ore.

The following assays must satisfy the most sceptical of the immense value of this property.

These results have been arrived at by several different scientific men, working entirely independent of each other.

No. I.—*Assay by Mr. John Williams, Agent Portage Lake Copper Smelting Works, Lake Superior.*

P. L. S. Works, Oct. 6th, 1863.

JOHN MARTIN, ESQ.

I have been working the whole night, after you and Mr. Sheldon left me, making an assay of the Galena lead ore you left me, two or three weeks since, and which you said you took from the Eldorado mines. It yielded 2.40 per cent. of silver,

---

* T. Sterry Hunt, in *American Jour. Science*, No. 93, 1761.

which is nearly the same result obtained when I tried it before, the same day that you gave it to me.

Your obedient servant,

(Signed) JOHN WILLIAMS.

The above is at the rate of 48 lbs. per ton of ore.

No. II.—*Assay of Julius G. Pohle, Analytical Chemist.*

New York, Oct. 14*th* 1863.

I have analyzed a sample of Argentiferous Galena for D. Allerton, Esq. It was found to yield, by fire assay, 45.66 per cent. of lead.

The lead yielded silver in the proportion of 223 ounces (Troy) to the ton of 2,000 lbs., avoirdupois.

(Signed) JULIUS G. POHLE,
Late of Jas. R. Chilton & Co.,
Analyitical Chemist.

Assay No. III.—*By the same.*

New York, Oct. 14*th*, 1863.

I have analysed a sample of Argentiferous Galena for A. H. Sibley.

It was found to yield 65 per cent. of lead. The lead yielded silver in the proportion of 148 ozs. 1 dwt. and 16 grains, to the ton of 2,000 lbs.

(Signed) JULIUS G. POHLE.
Late of Jas. R. Chilton & Co.,
Analytical Chemist.

Assay No. IV.—*By Professor Douglass, of the University of Michigan.*

Ann Arbor, Oct. 26*th* 1863.

R. Shelden, Esq., Houghton.

Dear Sir:—Yours of October 5th was duly received; I have this moment completed the analysis of the Galena received by express; it contains 3.26 per cent. of pure silver, or upwards of sixty-five pounds of silver to the ton of ore; I suppose that I took an average of the specimen, yet the yield is very large; I send you the result, with a sample of the silver extracted; I have worked nights to complete the analysis in order to send it early; I will make a second analysis.

Yours, truly,

S. H. DOUGLASS.

Assay No. VI.—*By Dr. Torrey, of the Assay Office, N. Y.*

United States Assay Office,
New York, Nov. 9th, 1863.

A. H. Sibley, Esq.

Dear Sir:—Your three small lumps of ore yield silver at the rate of \$118 $\frac{22}{100}$ per ton of 2000 lbs.

Respectfully,

(Signed) JOHN TORREY,
Assayer.

Assay No. V.—*By A. A. Hayes, of Boston, Mass.*

Results of assay of galena in quartz received from C. M. Sanderson, Nov. 1st, 1863.

There is nothing peculiar in the appearance of this ore; but testing shows it to be rich in silver.

The assay was made on rock and ore without dressing it.

One hundred parts contain 37 $\frac{22}{100}$ of silver-lead, a little iron and sulphur united with the lead.

When *all* the lead in the ore has been reduced to the metallic state, it contains 145 oz. 4 dwt. and 1 gr. of pure silver in each ton of lead; but as smelting operations give less lead than the ore contains, more silver per ton of lead will be found in the large operations.

Respectfully,

A. A. HAYES, M. D.,
State Assayer,
16 Boyleston St., Boston, Mass.

The above assays were made from ore taken out of Sections 6 and 36 of the Smith vein.

The following assays were made from ore taken from a vein passing through Section 6 :

ASSAY NO. VII.—*By Professor Cassels, of the Medical College, Cleveland, Ohio.*

Cleveland, November, 19, 1863.

ROBERT HANNA, Esq.

Dear Sir,—The specimens of lead ore you sent me for analysis, I found to consist of—

Galena	98.03
Oxide of Iron	0.55
Silver	1.42
	100.00

The above percentage of silver is equivalent to 28.40 lbs. of silver to the *ton of ore.*

Yours, truly,

J. L. CASSELS.

ASSAY NO. VIII.—*By Professor A. DuBois (late of the University of Michigan) and C. P. Williams, Analytical Chemist, Philadelphia.*

LABORATORY, No. 138 Walnut St.,
Philadelphia, Dec. 3d, 1863.

Messrs. HANNA & SPAULDING, Continental Hotel.

Gentlemen,—We have examined the specimens marked "Silver-lead from Section 14, Town 49 north, Range 28 West, Lake Superior," which you handed us, with the following results:

The ore was a coarsely crystalized galena contained in a gangue of quartz, with which is associated a small amount of a chloratoid mineral.

Sample No. 1 gave 8.480 per cent. lead, and .090 per cent.

silver; or, 169.60 lbs. lead *per ton of ore*, and 1.80 lbs. silver *per ton of ore*. In this sample the amount of silver per ton of *lead* would be 21.80 lbs.

Sample No. 2 gave 11.060 per cent. lead, and .106 per cent. silver; or, 221.20 lbs. lead, and 2.12 lbs. silver *per ton of ore*. The amount of silver per ton of lead would be 18.53 lbs.

Respectfully yours,

DU BOIS & WILLIAMS.

The following statement will show at a glance the results arrived at by the different assayers and chemists, both as regards the amount of silver and its value. The calculations are made with lead at 8 cents per lb., and silver at the usual standard price. A cubic fathom of ground is taken at 20 tons:

ANALYST.	LBS. AVOIRDUPOIS PER TON OF ORE.		VALUE PER TON OF ORE.		TOTAL VALUE PER TON ORE.	VALUE PER CUBIC FATHOM.	REMARKS.
	LEAD.	SILVER.	LEAD.	SILVER.			
J. Williams	1740	48.	$139.00	$960.00	$1,099.00	$21,980.00	Nearly pure galena.
Hayes	744.4	3.4	59.55	68.00	127.55	2,551.00	Ore yielded 37.22.
Douglass	1740	65.	139.20	1,300.00	1,439.00	28,780.00	Pure galena.
Pohle...............	1106	5.8	88.48	116.00	214.48	4,289.60	Mean of two analyses.
Cassel	1700	28.4	136.00	56.80	292.80	5,856.00	Nearly pure galena.
Torrey	not est.	not est.	not est.	118.22		2,364.00	Silver alone.
Du Bois & Williams..	195.4	1.96	15.63	39.20	54.83	1,096.60	Mean of two analyses.
Average	1204.3	25.43	96.34	665.46	761.80	15,236.00	

That the value of this property may be the more clearly understood by those not conversant with mining enterprises, we submit the above table, showing the yield from the result of assays by the most scientific chemists in the United States.

The largest yield of silver per ton of lead, is $1,300.

The average yield, per ton, is $665.46.

The average yield, per fathom of ground, is $13,309 in silver. The lowest yield, per fathom, is $785.

By reference to the above table, it will be shown that the smallest yield in silver *and* lead is $1,096.60 per fathom.

We do not wish to convey the impression that the yield will be equal to the average shown. If it will yield one-sixtieth of the average above shown, or less than one-fourth of the lowest yield arrived at, say $250 per fathom, the mine will pay a profit of *several millions of dollars per annum.*

The following statistics, taken from the report of the Quincy Mining Company, for the year 1862, will show the cost of mining 4,613 fathoms of ground.

The total cost (including all expenses) is...	$365,896.89
Cost per fathom........................	79.76
Yield per fathom........................	125.40
Profit "	46.05

If the silver and lead will yield the lowest result arrived at, it will show a profit equal to $4,687,000 from 4,613 fathoms of ground.

If we base our figures upon one-fourth of the lowest yield, or one-sixtieth of the average, we have a profit of $1,171,750 from 4,613 fathoms of ground worked.

It will be borne in mind that the territory of this

Company, now proposed to be formed, is sufficient in extent for the organization of ten different companies, each of which could mine the above amount of ground per annum.

We do not hesitate to assert that the yield in silver is greater, per ton of lead, than any other silver-lead mine ever worked. The only doubt which can arise in the mind of the most sceptical is as to the quantity of lead contained in the vein. The yield in silver cannot be disputed, the results above shown being conclusive upon that subject.

The following letter from Benjamin S. Lippincott, Esq., a gentleman who has visited and is well acquainted with the California and Washoe mining districts, will show that at the time he visited this property his only apprehension was as to the amount of silver contained in the lead.

New York, Dec. 10th, 1863.

A. H. Sibley, Esq.,

Dear Sir,—Agreeably to your request I submit the following as the result of my examination of the silver-lead mine known as the "Smith Vein:"

The mine is situated on a branch of the north fork of Dead River, about the center of Section 6, Town 49, Range 28, County of Marquette, State of Michigan.

The quartz is first seen in a perpendicular ledge of talcose slate and trap, the latter bearing a south-east and north-west course.

The quartz is a mixture of white and dark grey, and, to a practised eye in search of yet more precious metal, would lead to the conclusion, not only from the formation, but from the quartz itself, that the vein was auriferous. A heavy main lode of quartz carrying this argentiferous galena, is seen to run with the formation through the above mentioned ledge of slate and trap.

To a casual observer the veins of quartz will appear shattered and scattered, but this only occurs where it has been subjected to the action of water.

By following the branch above mentioned, you come abruptly upon a high wall of slate and trap, something in shape of a horse-shoe.

The walls of this ledge are traversed in every direction by veins or lodes of quartz, varrying in size, so that it is difficult to determine what dip predominates. At the base of this ledge an abundance of quartz detached and in boulders is found. The lode or main vein was traced about 500 yards to the south-east and about 400 yards to the north-west. Much trouble and labor was experienced in tracing the lodes, from the fact that there is an accumulation of moss and decayed vegetation, combined with the thick undergrowth covering this region.

At a distance of 150 yards north-east, and at an elevation of about 80 feet, there is a Beaver dam of four acres of water, fed by a lake some three-quarters of a mile still to the north.

From the dam and following the stream towards the mine, occur numerous cascades of three and four feet, until in the vicinity of the mine the water leaps down a declivity of 25 feet, affording a most excellent water power.

The hills in the immediate vicinity abound with most excellent timber for building and mining purposes.

At the time I visited the property, in September last, the only doubt in my mind was as to the amount of silver contained in the lead, and of which fact I had no means of judging. I unhesitatingly stated at that time that if the galena was as rich in silver as represented by those who claimed to have assayed it, I believed it to be only second to the great Washoe silver discoveries.

While I am satisfied that the most sanguine will not be disappointed in the yield of lead, I do not wish to be understood as asserting that very large dividends would be earned by working it as a lead mine, although it is my opinion that the yield of lead would earn a profit beyond the cost of working.

The assays of chemists of well known worth and standing,

shown me since I visited the property, have entirely removed the only apprehension I had at the time of my examination as above stated, and fully satisfy me that I did not over-estimate the immense value of this property.

Very truly yours,
B. S. LIPPINCOTT.

Especial attention is invited to the following letter from Mr. Charles P. Williams, an experienced mining geologist, late of the Lake Superior and Colorado mining regions, who unites to his experience an extensive and thorough acquaintance with the literature of mining. The statements made therein are grounded on well-established scientific facts, or substantiated by the best authors on mining, and were the result of a letter requesting his views on the chance of success in a section of country from the veins of which he examined several specimens.

LABORATORY, No. 138 Walnut Street,
Philadelphia, Dec. 5th, 1863.

JAMES CARSON, Esq., New York.

My Dear Sir,—In reply to your favor of the 4th inst., containing accounts of the new discoveries of silver-lead in the region of country on the south side of Lake Superior, between Keweenaw Bay and the Marquette Iron Region, and asking my opinion of the value of the results of the explorations and assays, I should say that, as far as my knowledge extends, I am very favorably impressed.

It would be beyond the compass of this letter to enter upon the discussion of a theory of the formation of these deposits of metallic sulphurets on these old rocks; but this much may be mentioned—that the most recent views of geologists point towards the hypothesis of their origin being due to "the action of organic matter which reduces sulphates, giving rise to metallic sulphides and sulphur." This view is supported by the discovery of obscure traces of organic remains in these rocks, which were formerly supposed to have been deposited at a period anterior to the appearance of life on the globe.

I should judge, from the general appearance and crystalline structure of such specimens of vein-matter as I have seen, from the metalliferous occurrences of this new mining section, that these latter are of the class known as true or fissure-veins. If this opinion be well founded, then they have two primary elements of valuable deposits—mode of occurrence and age of the enclosing rock; for, it is well established by mining experience all the world over, that fissure-veins are the only class of metalliferous deposits that are to be relied on for permanent richness, and that in the older and more crystalline rocks the greater amounts of silver ores or of argentiferous galenas occur.

It is true that fissure-veins are liable to fluctuations in richness, but no case is on record which has shown a permanent decrease in yield in depth, and has not given, on the average, better results, the more profound the workings. This can be asserted with more positiveness for plombiferous deposits than for those of the ores of any other metal. It has been proved in districts where due perseverance has been applied. In the Harz many of the mines are from two hundred to three hundred fathoms in depth. At Andreasburg, in the same mountains, the vein of the Sampson shaft is still productive at four hundred and thirty fathoms from the surface.

At Pizibram, in Bohemia, the richest silver-lead mines of the present day in Europe are worked at various depths, extending to nearly three hundred fathoms. The veins are composed mainly of galena in quartz, with some heavy spar and brown spar. Since they never yielded lead ores shallower than forty fathoms from the surface, they were for a long time neglected, as too poor for working. From that depth, downwards, the richness of the galena in silver increases. There are thirteen veins in the district, comprised in rocks of the Lower Silurian age.

The only lead mines in Cardiganshire which have attained a considerable depth are Loganlas and Goginan; and these have well rewarded the attempt to break through a prejudice that the veins of the district are not productive for deep working. The history of the last of these is particularly instructive. Its condition appeared to be perfectly hopeless, when it was taken in hand by the Messrs. Taylor & Co. "The previous adventurer, whose excavations extended to a depth of thirty or forty fathoms, repeatedly assured them that it was in vain to expect anything there, for, after many years experience, he was so satisfied that no more ore remained in the lode, *that he would undertake to carry on his back to Aberystwith all that he would ever extract;* and yet, in spite of his predictions, the mine has for years produced upwards of 1,500 tons of silver-lead ore per annum,"* paying an annual profit of $42,000 on a capital of only $2,500.

It would be needless to multiply examples in illustration of the permanent value of true fissure-lodes in depth, beyond the few cases above given, since this view has universally been received as a fixed principle by geologists, and has served for years as a guide in mining operations.

You have, on the property of which you wrote, a vein which satisfies this condition, which is situated in the most highly metalliferous of rocks, and which, on the authority of the reports of reliable engineers and explorers, is of good width, and exhibits, at its outcrops, and where it has been examined in more detail, a most excellent show of galena contained in a very favorable matrix. These facts, in connection with what has been substantiated above, are sufficient to warrant the inauguration of mining operations, with good prospects of successful results.

These prospects are greatly enhanced by the well-established fact that the galena is rich in silver. The assays (of which you furnished me copies) show a tenure of silver which is really wonderful. Of the great richness of the ore in the precious metal, I am fully satisfied, by analysis made by Prof. DuBois and myself of specimens from a vein in this district, the results of which I am at liberty to communicate to you. Two samples of the *undressed ore*, with a large amount of vein-stone, principally quartz, gave respectively 1.80 lbs. and 2.12 lbs. silver per ton, or at the rate, estimating the lead, of upwards of twenty lbs. per ton of metal.

* Mem, Geol. Survey, Great Britain, 22, p. 671.

You are aware that by the process now very generally adopted in England and elsewhere, a tenure of silver equal to six ounces per ton of lead will cover all expenses attending the separation of these two metals.

The most remarkable center of production of silver-lead ores in this country is at Washoe, at the eastern foot-hills of the Sierra Nevada, in latitude 39° N. "Here are powerful veins, ledges, and masses of quartz, that carry lead and silver. The silver is frequently seen in its native form in thread-like coils, but is generally diffused throughout galena or sulphuret of lead." This region has produced, during 1862, according to the San Francisco *Price Current*, $6,000,000, principally in silver.

In any generalizations upon the table of assays which you have prepared, it is only proper to disregard these analysis which were made on specimens which were almost entirely galena, and in our calculations only to take into account those with the adherent vein-stone, since they will represent more clearly the true yield of the vein. Hence, to be on the safe side, the last, which yielded only about 12 per cent. of galena and 88 per cent. gangue, will be selected. The total value of a cubic fathom of ground weighing twenty tons, with lead at 8 cents and silver at $20 per lb. (avoirdupois), on the above data will be $1,096.60. The cost of mining the same amount (when the mine shall have been opened), inclusive of dressing, need not exceed $45. This is somewhat more than now obtains among the copper mines of Lake Superior.

In our consideration of the economies of this new region we must add to the above cost of smelting the *dressed ores* and that of the separation of the contained silver from the resulting lead. Moissenet, in his *Notice Sur les Usines a Plomb de Pontesford* (Ann. des Mines, Sept. 1860), gives the following as the specifications of the costs of treatment of lead ores at Snailbeach, Shropshire, England.

	Per Ton Ore.	Per Ton Lead obtained
Workmen's wages....	$1 01	$1 40
Coal................	2 28	2 90
Sundries............	18	25
Total............	$3 47	$4 55

The ores treated here are those mined in the neighborhood of Shrewsbury, and are concentrated on the dressing floors to 71.87 per cent. Allowing 50 per cent. additional for the difference in wages between the two countries, we have about $4.00 as the expenses of treating one ton of 72 per cent. ore, or at the rate of $16.00 per cubic fathom of vein matter when concentrated to the above per centage. Or, in the aggregate, expenses will be as under:

Mining, crushing and dressing.....	$45	per cubic fathom.
Smelting.........................	16	" " "
Separation of silver from lead.....	10	" " "
Total expenses..............	$71	" " "

Exclusive of office charges, superintendence, etc. These calculations are of course only approximate, but they are made on liberal bases and with a low allowance for the yield of the vein-stuff.

If, then, sufficient exploration has been made to demonstrate the existence of good-sized veins, of which you are the best judge, having the reports of reliable gentlemen in the Lake Superior region, the chances for profitable mining operations are sufficiently great to satisfy the most sanguine adventurer.

Wishing you all success in your efforts towards the development of the vast metallic richness of the Lake Superior region, and with which you have been so long identified, and with the hope that the above data may be of assistance to you,

I remain, my dear sir,

Yours, most respectfully,

CHAS. P. WILLIAMS,

Chemist and Mining Geologist.

From the foregoing letter, in connection with the annexed copies of the results of assays of the ore from the property of the Silver and Lead Mining Company of Lake Superior, the conclusion is inevitable that this location is one of great value, and it is confidently believed that this discovery is second only in value to the far-famed Washoe mines. A comparative moderate outlay in mining expenses must insure enormous returns for the money invested.

The proprietors of these lands, therefore, feel satisfied that the chances of success are of the highest order, and in this belief invite the attention of capitalists to the foregoing statements, that they may assist in the legitimate development of a new field for mining enterprise in a new region, the great metallic wealth of which is already fully established. The title to the property, it may be well to remark, is in *fee-simple.*

www.ingramcontent.com/pod-product-compliance
Lightning Source LLC
LaVergne TN
LVHW020634110826
845149LV00004B/1180

* 9 7 8 1 4 1 8 1 9 1 5 9 7 *